FUN FACT FILE: HABITATS

20 FUN FACTS ABOUT DESERT HABITATS

BY JILL KEPPELER

Gareth Stevens PUBLISHING

Please visit our website, www.garethstevens.com. For a free color catalog of all our high-quality books, call toll free 1-800-542-2595 or fax 1-877-542-2596.

Library of Congress Cataloging-in-Publication Data

Names: Keppeler, Jill, author.
Title: 20 fun facts about desert habitats / Jill Keppeler.
Other titles: Twenty fun facts about desert habitats
Description: New York : Gareth Stevens Publishing, 2022. | Series: Fun fact File: Habitats | Includes index.
Identifiers: LCCN 2020036346 | ISBN 9781538264393 (library binding) | ISBN 9781538264379 (paperback) | ISBN 9781538264386 (6 pack) | ISBN 9781538264409 (ebook)
Subjects: LCSH: Desert ecology–Juvenile literature.
Classification: LCC QH541.5.D4 K46 2022 | DDC 577.54–dc23
LC record available at https://lccn.loc.gov/2020036346

First Edition

Published in 2022 by
Gareth Stevens Publishing
29 East 21st Street
New York, NY 10010

Designer: Michael Flynn
Editor: Kate Mikoley

Photo credits: Cover, p. 1 (main) Westend61/Getty Images; file folder used throughout David Smart/Shutterstock.com; binder clip used throughout luckyraccoon/Shutterstock.com; wood grain background used throughout ARENA Creative/Shutterstock.com; p. 5 Diana Robinson Photography/Moment/Getty Images; p. 6 José Carlos Diaz Hidalgo/EyeEm/Getty Images; p. 7 NicoElNino/iStock/Getty Images; p. 8 danm/Moment/Getty Images; p. 9 xingmin07/E+/Getty Images; p. 10 Oli Anderson/Moment/Getty Images; p. 11 Jeff Foott/Photodisc/Getty Images; pp. 12, 13 (kangaroo rat) C. Allan Morgan/Photolibrary/Getty Images; p. 13 (desert tortoise) Scott Smith/Corbis Documentary/Getty Images; p. 13 (thorny desert lizard) Floriane Mangiarotti/EyeEm/Getty Images; p. 13 (African lungfish) feathercollector/iStock/Getty Images; p. 13 (water holding frog) https://en.wikipedia.org/wiki/Ranoidea_platycephala#/media/File:Cyclorana_platycephala.jpg; p. 14 Eric Lo/Moment/Getty Images; p. 15 Hannes Thirion/Moment/Getty Images; p. 16 Timothy Allen/Stone/Getty Images; p. 17 Wolfgang_Steiner/E+/Getty Images; p. 18 Franz Aberham/The Image Bank/Getty Images; p. 19 cinoby/E+/Getty Images; p. 20 Douglas Sacha/Moment/Getty Images; p. 21 Julien Viry/iStock/Getty Images; p. 22 Cavan Images/Cavan/Getty Images; p. 23 Mitchell Krog/Moment Open/Getty Images; p. 24 Joe McDonald/The Image Bank/Getty Images; p. 25 hagit berkovich/Shutterstock.com; p. 26 Tier Und Naturfotografie J und C Sohns/Photographer's Choice/Getty Images; p. 27 Shongololo90/iStock/Getty Images; p. 29 ericfoltz/E+/Getty Images.

Printed in the United States of America

CPSIA compliance information: Batch #CSGS22: For further information contact Gareth Stevens, New York, New York at 1-800-542-2595.

CONTENTS

Words in the glossary appear in **bold** type the first time they are used in the text.

DRY DESERT HABITATS

What makes a desert? These important **habitats** cover about 20 percent of Earth's surface. They're arid, or very dry, and get less than 10 inches (25.4 cm) of rain each year. Because of this, deserts support limited amounts of plants and animals.

However, some life-forms have **adapted** to live quite well in deserts. Plants such as cacti and a number of animals, including many birds and **reptiles**, call this habitat home. If you look closer, you'll find lots of interesting things in a desert!

Even in deserts, plants and animals find ways to survive. Few parts of a desert are really lifeless.

WHERE ARE DESERTS?

FUN FACT: 1

DESERTS ARE FOUND ALL OVER THE WORLD!

The **continents** of Africa and Australia are known for their many desert habitats, but even Europe has at least one. The Tabernas Desert in Spain is the small continent's only true desert.

The Tabernas Desert looks so much like the U.S. Southwest that movie studios sometimes use it to shoot movies set in that area.

Antarctica (and its desert) covers about 5.5 million square miles (14.2 million sq km). There's not much plant or animal life on land, but the waters around it are home to many creatures.

FUN FACT: 2

ONE CONTINENT IS *ALL* DESERT.

Icy Antarctica is a desert. Like a hot desert, it gets very little rainfall, although it's covered in snow and ice. This is called a polar desert. One covers much of the Arctic region too.

Many years ago, the Sahara was a green grassland. Many early peoples likely lived there.

FUN FACT: 3

THE SAHARA IS THE BIGGEST HOT DESERT. MOST OF THE UNITED STATES COULD FIT INSIDE IT!

The Sahara Desert in Africa stretches across the northern part of the continent. It covers more than 3.5 million square miles (9.1 million sq km).

THE WORLD'S BIGGEST DESERTS

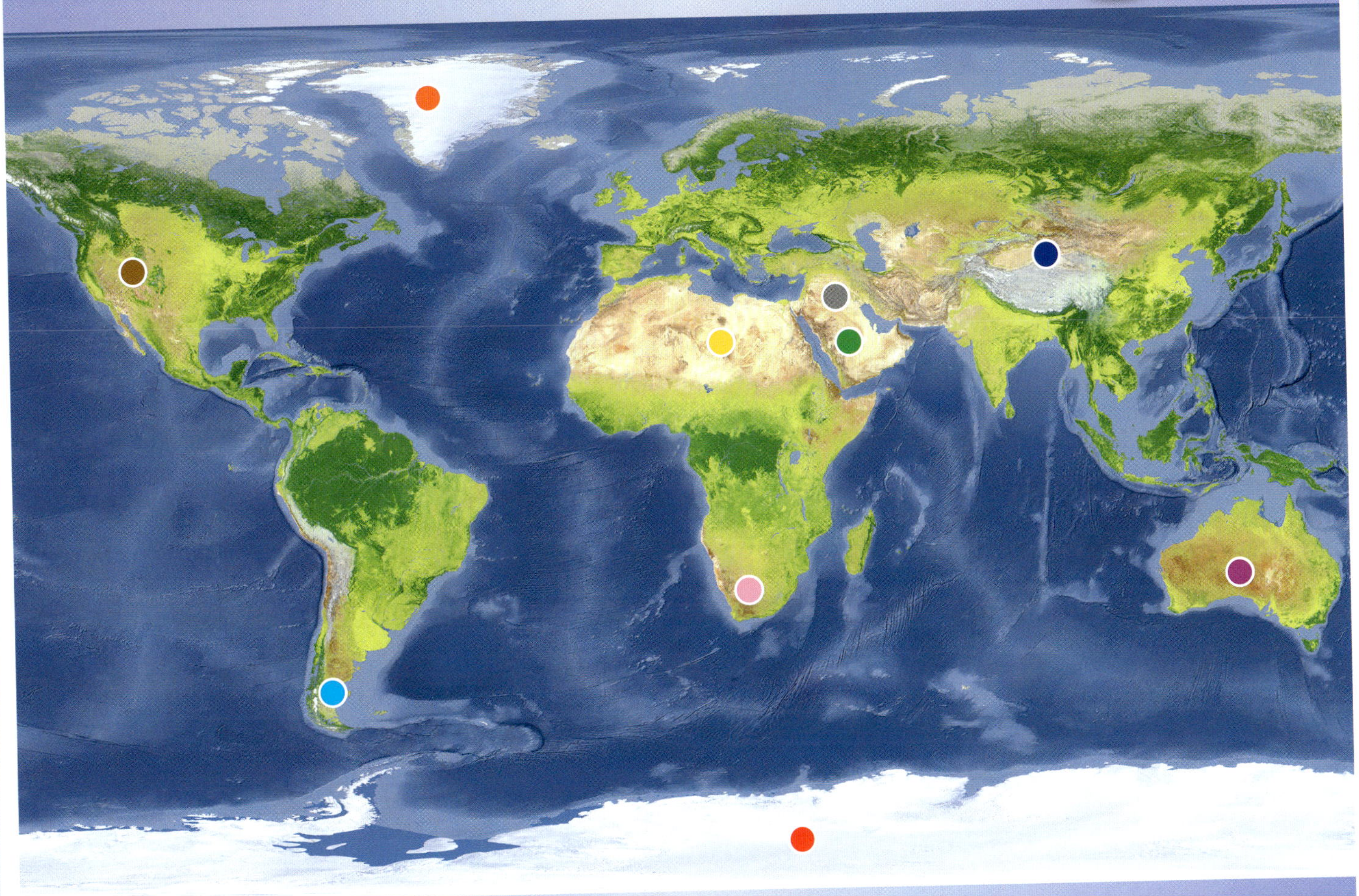

- ANTARCTIC
- ARCTIC
- SAHARA
- ARABIAN
- GOBI
- PATAGONIAN
- GREAT VICTORIA
- KALAHARI
- GREAT BASIN
- SYRIAN

Located mainly in Nevada and Utah, the Great Basin is the biggest desert in the United States.

REALLY RARE RAINFALL

FUN FACT: 4

A DESERT MAY GET NO RAINFALL AT ALL FOR YEARS!

Iquique, a city in Chile, had no rain at all for four years. Then, there was finally a rain shower—which brought 0.6 inches (1.5 cm) of rain to the area.

Iquique is located in South America's Atacama Desert. It's one of the driest places in the world.

Virga can be very beautiful. It can appear like cloudy threads in the air.

FUN FACT: 5

IN THE DESERT, RAIN CAN SOMETIMES EVAPORATE BEFORE IT HITS THE GROUND.

Precipitation that evaporates before it reaches the ground is called virga. This doesn't just happen in deserts, however. It just needs dry air and high temperatures.

A kangaroo rat has many adaptations that help it **conserve** water from seeds it eats.

FUN FACT: 6

SOME DESERT CREATURES CAN LIVE FOR YEARS WITHOUT DRINKING WATER.

These animals include the kangaroo rat, the desert tortoise, the thorny devil, and the water-holding frog. The kangaroo rat can live its whole life without drinking water!

DESERT CREATURES AND WATER ADAPTATIONS

KANGAROO RAT

- CHEEK POUCHES
- SPECIAL BODY PARTS
- DOESN'T SWEAT

DESERT TORTOISE

- CAN STORE HUGE AMOUNTS OF WATER IN ITS BODY—UP TO TWO-FIFTHS ITS WEIGHT!

THORNY DEVIL

- SCALES CAPTURE RAINFALL AND DEW

WATER-HOLDING FROG

- DRAWS IN WATER THROUGH ITS SKIN AND STORES IT
- CAN BURY ITSELF IN THE GROUND AND LIVE OFF ITS SKIN DURING DRY SEASONS

WEST AFRICAN LUNGFISH

- GILLS CAN PULL OXYGEN FROM WATER OR AIR
- CAN LIVE IN DRIED-UP MUD

Living things need water to stay alive, but in deserts, some animals have adapted so they rarely have to drink any!

HOT, HOT, HOT

FUN FACT: 7

HOT DESERTS CAN BE *VERY* HOT—REGULARLY ABOVE 100°FAHRENHEIT (37.8°CELSIUS) DURING THE DAY!

The highest recorded temperature on Earth to date was 134°F (56.7°C) in Death Valley in 1913.

Death Valley lies at the border of the Great Basin and Mojave Deserts in the United States.

Most hot deserts, such as the Kalahari in southern Africa, are located about midway between Earth's poles and its **equator**.

FUN FACT: 8

HOWEVER, THE TEMPERATURE CAN DROP BELOW FREEZING AT NIGHT. BRRR!

Because of the lack of water in a desert, there are no clouds or water vapor in the sky to hold heat in at night. It can get very cold.

SO MANY EXTREMES

FUN FACT: 9

SOME KINDS OF DESERTS CAN HAVE VERY HOT SUMMERS—BUT VERY COLD WINTERS.

Cold deserts are usually farther above sea level, sometimes in the mountains. Temperatures there during winter can get far below freezing!

The Gobi Desert in Central Asia is a cold desert. Its winters can be as cold as −40°F (−40°C)!

The Namib Desert in Africa stretches about 1,200 miles (1,931.2 km) along the Atlantic coast.

FUN FACT: 10

DESERTS AND OCEANS CAN EXIST RIGHT NEXT TO EACH OTHER!

All that water—but none for the desert next door! Coastal deserts, located near where the land meets the ocean, often have very little rainfall, even less than other deserts.

DUNES AND OASES

FUN FACT: 11

MOST DESERTS PROBABLY DON'T LOOK THE WAY YOU'RE EXPECTING THEM TO LOOK!

Although movies often show deserts covered in sand dunes, these hills of sand piled up by wind cover only 10 percent of the world's deserts.

Sand dunes are a beautiful desert feature, but deserts can also be flat, rocky, or mountainous.

Oases can be as small as 2.5 acres (1 ha) or much larger.

FUN FACT: 12

EVEN IN THE DESERT, THERE ARE SPOTS OF GREENERY.

An oasis is a green, **fertile** area in the middle of a desert. Even the mighty Sahara has about 90 major oases! Their water often comes from underground springs.

COOL CACTI AND OTHER PLANTS

FUN FACT: 13

A SAGUARO CACTUS CAN TAKE IN AND STORE AS MUCH AS 200 GALLONS (757.1 L) OF WATER!

The desert plants called cacti soak up and store water in their cells. They can store it a long time.

A saguaro cactus, like the ones shown here, can sometimes grow to be almost 50 feet (15.2 m) tall.

Scientists aren't sure how the mistletoe cactus came to live in Africa. Birds or humans may have carried its seeds to the area.

FUN FACT: 14

YOU WON'T SEE A SAGUARO IN THE SAHARA!

Almost all cacti—including the saguaro—are only native to North and South America. Only the mistletoe cactus also lives in East Africa, Madagascar, and Sri Lanka.

Plants belonging to the daisy family can be found in deserts in North America, Australia, Africa, the Middle East, and more locations.

FUN FACT: 15

THE MOST DIVERSE DESERT PLANTS WORLDWIDE ARE MEMBERS OF THE DAISY FAMILY.

You probably don't think of flowers when you think of the desert! However, there are more than 23,000 species, or kinds, of plants in the Asteraceae, or daisy, family. Many live in deserts!

FUN FACT: 16

SOME DESERT PLANTS CAN BE VERY, VERY OLD!

The tumboa, or welwitschia, plant lives in the Namib Desert in Africa. It only has two leaves, and some plants may be 2,000 years old.

One ring of Mojave yucca plants in the United States may be more than 12,000 years old! Creosote bushes in the Mojave Desert may also be about that old.

CREATURES OF THE DESERT

FUN FACT: 17

DESERT CREATURES HAVE MANY UNUSUAL ADAPTATIONS. SIDEWINDER SNAKES HAVE A SPECIAL WAY OF CRAWLING ACROSS HOT SAND!

Sidewinders, which live in North America, Africa, and the Middle East, move with only two points of their body in contact with the sand.

A sidewinder moves sideways across the ground, leaving an interesting trail.

Fennec foxes only weigh about 2.2 pounds (1 kg).

FUN FACT: 18

THE TINY FENNEC FOX, WHICH LIVES IN NORTH AFRICA, HAS BIG EARS THAT HELP IT KEEP COOL.

These ears, full of blood **vessels**, give off heat during hot desert days. They also help the fox listen for tiny bugs to eat.

FUN FACT: 19

KANGAROOS TAKE SPIT BATHS TO HELP THEM STAY COOL!

Desert-dwelling kangaroos have a special web of tiny blood vessels very close to the skin in their front legs. They lick their legs to cool themselves down.

Kangaroos live in Australia. A red kangaroo's body is about 70 percent water!

Meerkats live in African deserts. Their eye markings help them keep watch!

FUN FACT: 20

MEERKATS AND FOOTBALL PLAYERS HAVE SOMETHING IN COMMON!

Meerkats have black circles around their eyes—much like the "eye black" paint some football players wear in games. For both, the black marks keep the bright sunlight from shining into their eyes!

GROWING DESERTS

Many plants and animals call Earth's desert habitats home. However, some deserts are spreading—and that's not a good thing. Desertification is when land in already dry areas becomes drier and less likely to support life. This is often caused by human activities, such as overfarming and cutting down forests.

People are searching for ways to fight desertification, such as planting new trees. We need to both fight desertification and protect existing desert habitats—and all the plants and creatures that call them home.

Earth needs a balance between deserts and other habitats.

GLOSSARY

adapt: to change to suit conditions

conserve: to keep something from harm and not waste it

continent: one of Earth's seven great landmasses

diverse: differing from each other

equator: an imaginary line around Earth that is the same distance from the North and South Poles

evaporate: to change from a liquid to a gas

fertile: able to support many plants

gill: the body part that water animals such as fish use to breathe in water

habitat: the natural place where an animal or plant lives

precipitation: rain, snow, sleet, or hail

reptile: an animal covered with scales or plates that breathes air, has a backbone, and lays eggs, such as a turtle, snake, lizard, or crocodile

vessel: a tube that carries blood through the body

FOR MORE INFORMATION

BOOKS

Gardeski, Christina Mia. *All About Deserts.* North Mankato, MN: Capstone Press, 2018.

Perish, Patrick. *Fennec Foxes.* Minneapolis, MN: Bellwether Media, 2019.

Silverman, Buffy. *Let's Visit the Desert.* Minneapolis, MN: Lerner Publications, 2017.

WEBSITES

Desert Habitat
kids.nationalgeographic.com/explore/nature/habitats/desert/
Learn more about deserts, their creatures, and more through this National Geographic website.

Deserts
www.dkfindout.com/us/earth/deserts/
Quiz yourself on what you know about different kind of deserts, including their landforms and more.

Deserts
www.worldwildlife.org/habitats/deserts
Read more about desert species, why they matter, and how to help protect them.

INDEX